Mock Exams for Math C

The Best Tasks from Math Olympiads

(Volume 1)

Educational Collection *Magna-Scientia*

The Best Tasks from Math Olympiads

Mock Exams *for* Math Olympians

Volume 1

Michael Angel C. G., Editor

Title:
Mock Exams for Math Olympians. The Best Tasks from Math Olympiads. Volume 1.

Edition, Cover and Interior Design:
Michael Angel C. G.

Preface

It is well known that students who participate in Math Olympiads require a solid preparation both to acquire optimal mathematical skills as well as to manage their emotions, since in a competition like this the participants are subjected to pressure of time, anxiety, nervousness that can end up negatively affecting their concentration and therefore their performance during the competition. Therefore, it is evident that these students need a special preparation that involves not only the cognitive aspect but also the emotional aspect.

The present edition aims to achieve in the math Olympians the consolidation of their mathematical skills after successfully solving a group of mock exams containing a variety of carefully selected interesting problems, as well as giving them the confidence to successfully face the exams of any math competition. This educational material will be of great help to all students who participate each year in the main mathematics competitions for elementary and middle school in the United States and abroad; and in a very special way for those who are preparing for the MOEMS contest, whose exams have inspired this edition. Furthermore, the problems included herein are very similar to those proposed in the main elementary and middle school mathematics competitions in the United States such as MOEMS, Math Alpha Contest, Noetic Math Contest, Math Kangaroo in USA, etc.

This edition consists of a series of workbooks that bring together a collection of select problems by means of Mock Exams and is aimed at elementary and middle school students. Many of the problems included here have been extracted from Math Olympiads around the world and others have been inspired by them, which will allow the student to prepare by performing simulations of a math competition. Likewise, it has been considered to follow the structure and rules of the exams given in the MOEMS contests (Mathematical Olympiads for Elementary and Middle Schools) due to its great popularity in the United States and abroad. Furthermore, each Mock Exam contains 5 questions in increasing order of difficulty to be answered in a time not exceeding 30 minutes, where each correct answer is worth one point and the incorrect answer zero points. The main topics covered by the questions include: sets of numbers, arithmetic operations, math and logic puzzles, divisibility, prime numbers, GCF – LCM, fractions, statistics and probability, geometry in the plane and solids.

The exams included in each volume have been divided into two categories, namely, elementary school and middle school, each of them with a total of ten Mock Exams. In this first volume the exams from 1 to 10 are included. The students may only have: pencil, eraser and sharpener. Blank sheets will not be required as the workbook has been designed so that the students can solve each question in the same workbook. No calculators, rulers, graph paper, or any other aid can be used. In addition, the students will find the answers to each question at the end of the book, so that they can verify their results obtained. Finally, the indispensable support of parents or an academic tutor is recommended so that they can guide the student in case of doubts, and the evaluation is carried out with the greatest objectivity and responsibility possible.

Sincerely,

The editor

Contents

Preface **5**

Elementary School **9**

Mock Exam 1 11

Mock Exam 2 21

Mock Exam 3 31

Mock Exam 4 41

Mock Exam 5 51

Mock Exam 6 61

Mock Exam 7 71

Mock Exam 8 81

Mock Exam 9 91

Mock Exam 10 101

Middle School **111**

Mock Exam 1 113

Mock Exam 2 123

Mock Exam 3 133

Mock Exam 4 ... 143

Mock Exam 5 ... 153

Mock Exam 6 ... 163

Mock Exam 7 ... 173

Mock Exam 8 ... 183

Mock Exam 9 ... 193

Mock Exam 10 ... 203

Answers ... 215

Mock Exams – Elementary School ... 217

Mock Exams – Middle School ... 219

Elementary School

Mock Exam 1

(Elementary School)

Name: ______________________________ Grade: ________

Date: ____________ Start Time: ________ End Time: ________

Problem 1.

Calculate $12 + 39 + 456 + 88 + 44 + 261 + 11$.

Indicate the simplest order of actions.

Problem 2.

At my birthday party we were less than 10. A bag of candies was distributed so that each of us had 12 and 8 left over. How many candies were there?

Problem 3.

The three members of a rabbit family ate a total of 73 carrots. The father ate 5 more carrots than the mother. The son ate 12 carrots. How many carrots did the mother eat?

Problem 4.

Three erasers, one pencil and two notebooks cost 22 coins. One eraser, three pencils and two notebooks cost 38 coins. How many coins does a set of one eraser, one pencil and one notebook cost?

Problem 5.

A certain drink is prepared correctly if it is obtained by mixing 1 part of syrup with 5 parts of water. By mistake Albert mixed 5 parts of syrup with 1 of water, obtaining 3 liters of mixture. By adding an appropriate amount of water, Albert can obtain a drink in which the established proportions are respected. How many liters of water does he just need to add?

Answer Sheet

	Comment	Answer
1		
2		
3		
4		
5		

Score

1	2	3	4	5	Total

Mock Exam 2
(Elementary School)

Name: ______________________________ Grade: __________

Date: ______________ Start Time: __________ End Time: __________

Problem 1.

Nick and Sam run on the stadium track. Nick takes 3 minutes for each lap, while Sam takes 4 minutes for each lap. They leave at the same time. After how many minutes will they still cross the starting line together?

Problem 2.

A sheet of paper is folded five times as shown in the figure. A hole is then made in the center of the folded sheet. How many holes are there now on the unfolded sheet?

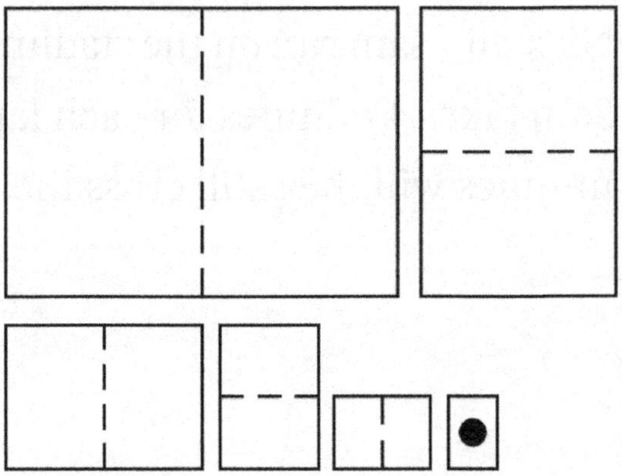

Problem 3.

In the following puzzle: $A + A + BB = CCC$, different letters correspond to different digits. Determine $A + B + C$.

Problem 4.

If the sum of five consecutive integers greater than zero is 3210. What is the largest number?

Problem 5.
The sum of two positive integers is 77. The smaller of the two multiplied by 8 gives the same result as the other multiplied by 6. What is the larger of the two numbers?

Answer Sheet

	Comment	Answer
1		
2		
3		
4		
5		

Score

1	2	3	4	5	Total

Mock Exam 3
(Elementary School)

Name: ______________________________ Grade: ________

Date: ____________ Start Time: ________ End Time: ________

Problem 1.

Calculate $422 - 67 + 446 - 38 + 187 - 12 + 78 - 36$.

Indicate the simplest order of actions.

Problem 2.

The length of a rectangular field is 80 m and the area is 3 200 m^2. Find the length of another rectangular field whose area and width are both half of the corresponding ones in the first field.

Answer Sheet

	Comment	Answer
1		
2		
3		
4		
5		

Score

1	2	3	4	5	Total

Mock Exam 4
(Elementary School)

Name: ______________________________ Grade: ________

Date: ____________ Start Time: ________ End Time: ________

Problem 1.

The large square is divided into small squares. What fraction of the unshaded area represents the shaded area?

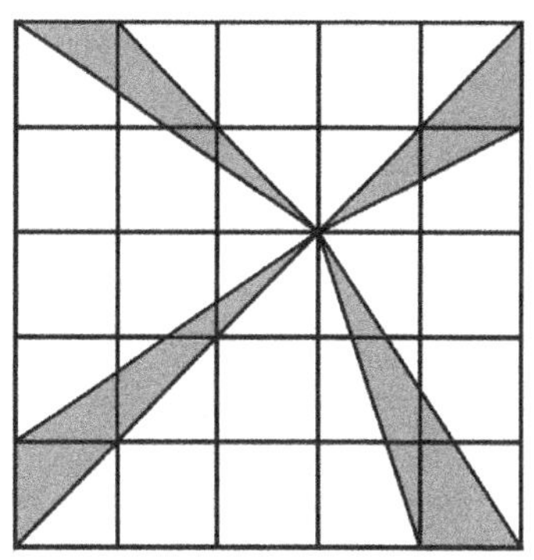

Problem 2.

Steven and Simon went to look for mushrooms in the forest, collecting a total of 70. In addition, 5/9 of the mushrooms collected by Steven are porcini and 2/17 of the mushrooms collected by Simon are oyster. How many mushrooms did Steven collect?

Problem 3.

In the figure 9 boxes are aligned; in the first appears the number 7 and in the last the number 6. What number should we write in the second box, if we want the sum of the numbers for each three consecutive cells to be 21?

7								6

Problem 4.

The figure shows some pearls (represented by little circles) connected to each other by threads (represented by segments). How many of these threads do you just need to remove to get a necklace, made up of a single ring that contains all the pearls?

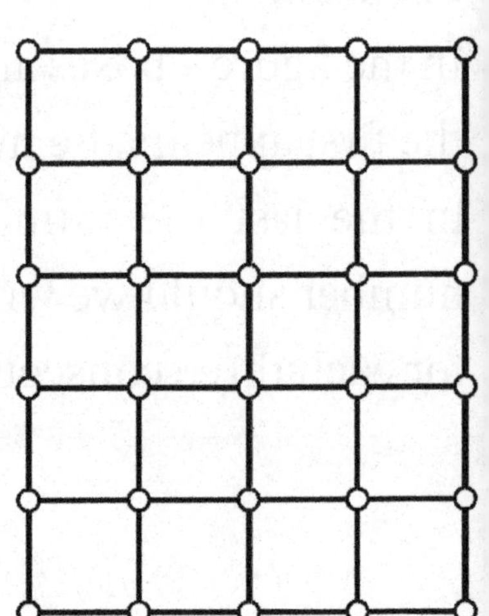

Problem 5.
Two items in a store had the same price a month ago. Subsequently, one of the two has had a 5% discount, while the other has experienced a 15% increase. According to these changes, the two prices now differ by 6 dollars. What is the current price of the cheapest item?

Answer Sheet

	Comment	Answer
1		
2		
3		
4		
5		

Score

1	2	3	4	5	Total

Mock Exam 5

(Elementary School)

Name: ______________________________ Grade: ________

Date: ____________ Start Time: ________ End Time: ________

Problem 1.

From the beginning of 2005 to the end of 2025, how many months are there that start and end on the same day of the week?

Problem 2.

The figure on the side is made up of five equal isosceles right-angled triangles. Find the area of the shaded part in square centimeters.

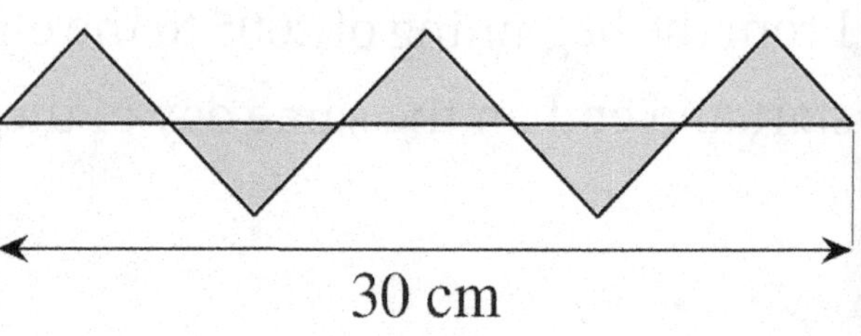

Problem 3.

Anya has 9 pencils in a box. There is at least one blue. In each group of four pencils taken out of the box, there are at least two of the same color. And, in each group of five pencils taken out of the box, there are at most three of the same color. How many blue pencils are there?

Problem 4.

By dividing the number 111 ... 11 (2004 digits equal to 1) by 3. How many zeros appear in the quotient?

Problem 5.

In the following cryptarithm: AAA + AAB + ACC = 2003, different letters represent different digits. Find the value of A^{B^C}.

Answer Sheet

	Comment	Answer
1		
2		
3		
4		
5		

Score

1	2	3	4	5	Total

Mock Exam 6
(Elementary School)

Name: ______________________________ Grade: ________

Date: ____________ Start Time: ________ End Time: ________

Problem 1.

Calculate:

$$\frac{2}{5}+\frac{22}{55}+\frac{222}{555}+\frac{2222}{5555}+\frac{22222}{55555}$$

Problem 2.

A square with side "x" is shown in the figure beside. Find the value of "x" in centimeters.

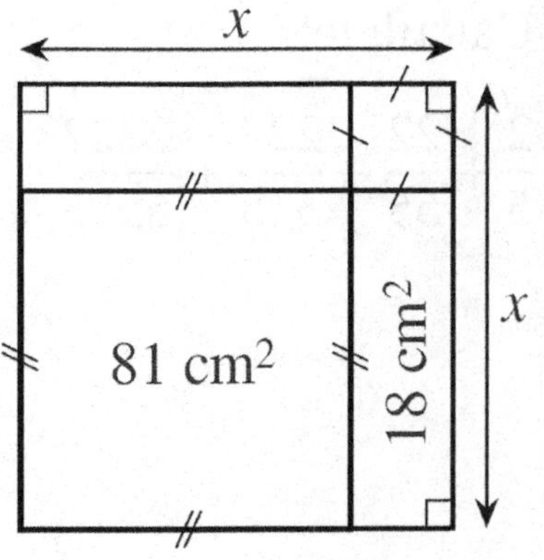

Problem 3.

The clown Benny is dancing, perched on two balls and a cubic box. The bottom ball has a radius of 6 *dm*; the top one has a radius three times smaller. The side of the cubic box is 4 *dm* larger than the diameter of the small ball. How high above the ground is Benny (in decimeters)?

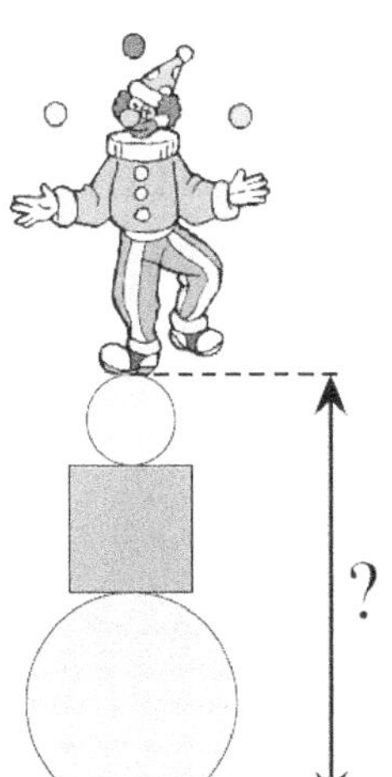

Problem 4.

The average age of the players of a soccer team on the field (11 in number) at the start of a match is 23 years. At the start of the second half, two players, both 26 years old, are replaced by a player of 20 and a player of 21. After these substitutions, what is the new average age of the team?

Problem 5.

Cut the corners of a square of cardboard and fold it into a box, as shown in the figure. What is the volume of the box in cubic centimeters?

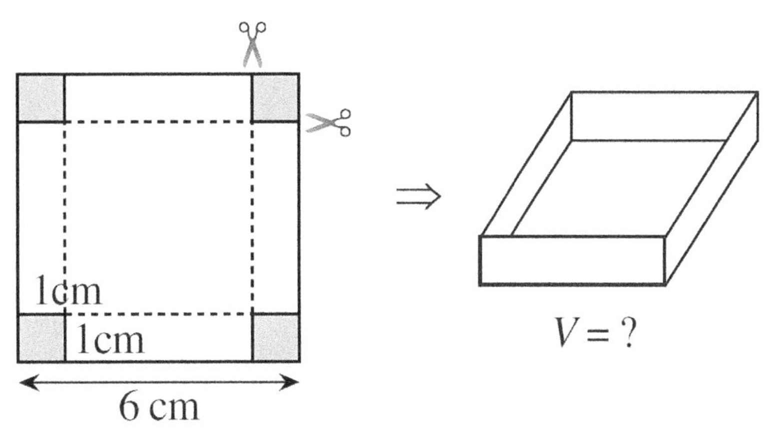

Answer Sheet

	Comment	Answer
1		
2		
3		
4		
5		

Score

1	2	3	4	5	Total

Mock Exam 7
(Elementary School)

Name: ______________________________ Grade: __________

Date: ______________ Start Time: __________ End Time: __________

Problem 1.

Find the value of “N” in the following expression:

$5 \times N = 2000 + 2001 + 2002 + 2003 + 2004 + 2005.$

Problem 2.

Consider only integers greater than 0. What is the difference between the sum of the first 1000 even numbers and the sum of the first 1000 odd numbers?

Problem 3.

A die is in the position shown in the figure. How many complete laps around the track will it take for the die to return exactly to its starting position, rotating one face at a time?

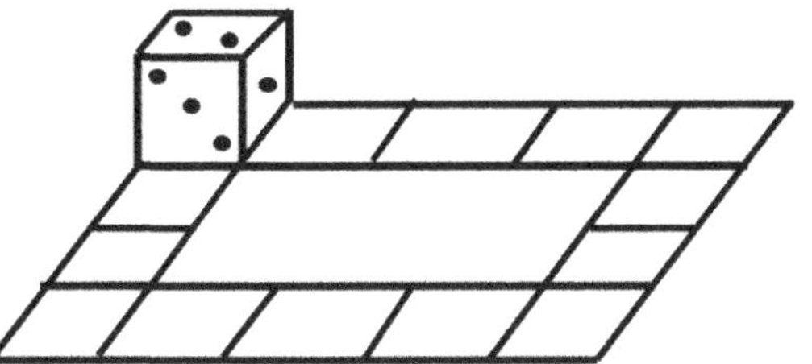

Problem 4.

Tony has 20 balls of different colors: yellow, green, blue and black. 17 balls are not green, 5 are black, 12 are not yellow. How many blue balls are there?

Problem 5.

There are 46 trees along the way from Liam's house to the pool. Going from the house to the pool and back, Liam has marked some trees with a ribbon as follows. On the way he marked the first tree and then the second of each pair of trees he encountered; on the way back, instead, he marked the first tree and then the third of each triad of trees that he encountered. After that, how many trees have the ribbon?

Answer Sheet

	Comment	Answer
1		
2		
3		
4		
5		

Score

1	2	3	4	5	Total

Mock Exam 8
(Elementary School)

Name: ______________________________ Grade: ________

Date: ____________ Start Time: ________ End Time: ________

Problem 1.

The square STUV is made up of an interior square surrounded by 4 identical rectangles. The perimeter of each of the rectangles is 40 *cm*. What is the area, in square centimeters, of the square STUV?

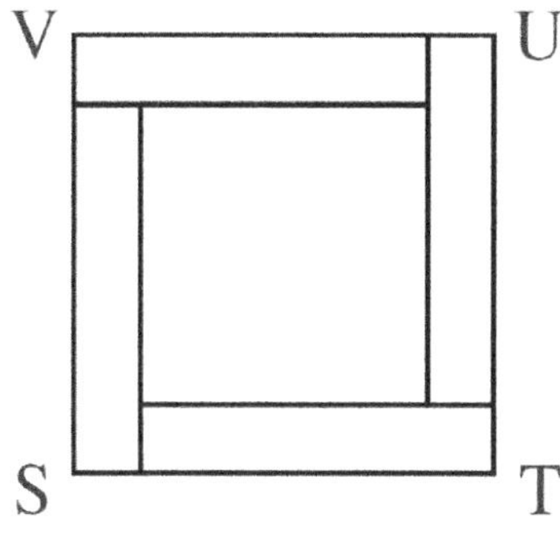

Problem 2.

Mick has a cube and wants to color three sides in blue and three in red. How many different ways can he do this? (Two cubes are meant to be colored differently if, no matter how one is rotated, the other cannot be obtained.)

Problem 3.

Four snails crawled across a floor made up of identical rectangular tiles. The figure shows the trace left by each of them. It is known that the trace left by Fin is 25 decimeters long, the one left by Pin is 37 decimeters long while the one left by Rin is 38 decimeters long. How many decimeters is the trace left by the snail Tin long?

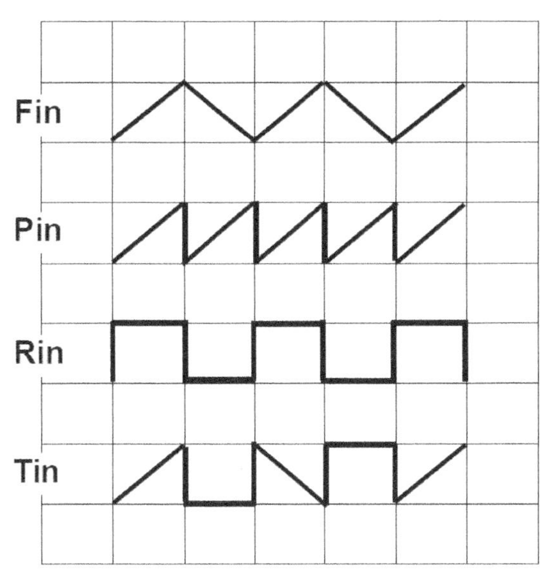

Problem 4.

Bruno had a 15 meter long rope. By cutting it appropriately, he obtained pieces of it having a whole number of meters in length. If the number of pieces is the maximum possible, and there are no two pieces of the same length. How many cuts has Bruno made?

Problem 5.

A segment OE is 2006 meters long. On it we identify three points A, B, C so that the segments OA and BE measure 1111 m and the length of OC is 70% of that of OE. Starting from the end O, in what order are the five points located?

Answer Sheet

	Comment	Answer
1		
2		
3		
4		
5		

Score

1	2	3	4	5	Total

Mock Exam 9

(Elementary School)

Name: ______________________________ Grade: ________

Date: ____________ Start Time: ________ End Time: ________

Problem 1.

A cylindrical glass with height 10 *cm* contains water. The figure shows it in two different positions. What is the height "x" (in centimeters) of the water when the glass is upright?

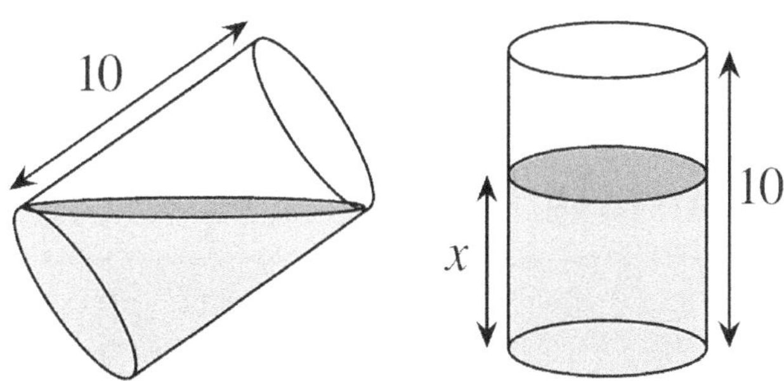

Problem 2.

In a game you have to count from 1 to 100 and clap every time you encounter either an integer multiple of 3 or a number ending in 3. How many times do you have to clap?

Problem 3.

Ronny has six sticks with lengths 1*cm*, 2 *cm*, 3 *cm*, 2001 *cm*, 2002 *cm* and 2003 *cm*, respectively. He has to choose three of them and form a triangle (which does not reduce to a segment). How many different choices of three sticks can Ronny make?

Problem 4.

In a moat there are red dragons and green dragons. Each red dragon has 6 heads, 8 legs and 2 tails. Each green dragon has 8 heads, 6 legs and 4 tails. There are 44 tails of all dragons. The number of green legs is 6 less than the number of red heads. How many red dragons are there in that moat?

Problem 5.

What is the length (in meters) of the path from F to G as shown in the figure beside?

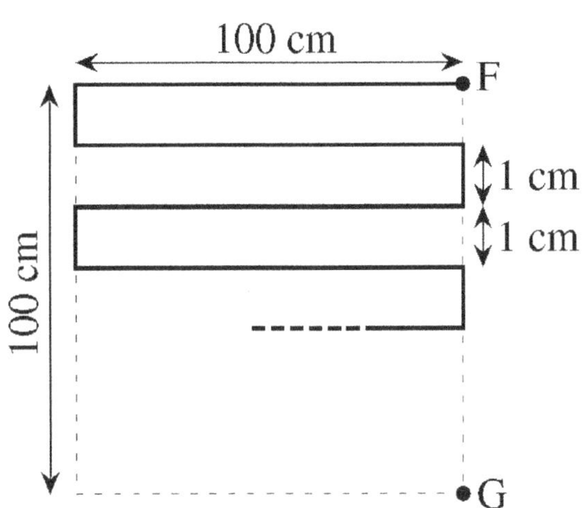

Answer Sheet

	Comment	Answer
1		
2		
3		
4		
5		

Score

1	2	3	4	5	Total

Mock Exam 10
(Elementary School)

Name: ______________________________ Grade: ________

Date: ____________ Start Time: ________ End Time: ________

Problem 1.

What is the smallest number of points that it is enough to remove from the figure shown on the side, so that there are not three remaining which are the vertices of some equilateral triangle?

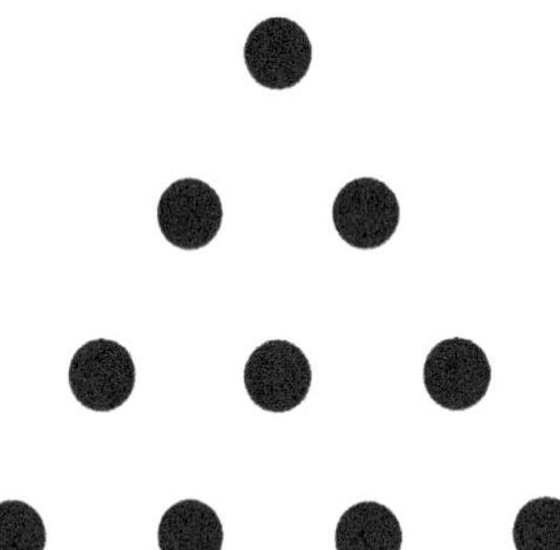

Problem 2.

In a campsite, Albert and Bryan prepare a bonfire to cook their food, using 15 identical pieces of wood; 8 were brought by Albert and 7 by Bryan. Carl asks them to use the same bonfire to cook, and rewards his friends with 30 coins (all of the same value). How many coins did Bryan receive if the distribution of the coins was done fairly?

Problem 3.

The star shown in the figure was made by touching the midpoints of the sides of a regular hexagon. If the area of the star is 12, what is the area of the hexagon?

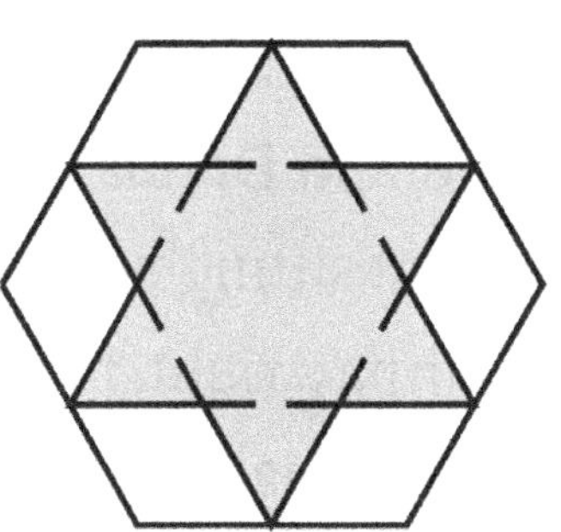

Problem 4.

Danna, Lana, Mary, Sara and Vera sit on a bench. Mary does not sit at the right end and Danna does not sit at the left end. Sara does not sit at either end. Vera is not sitting next to Sara and Sara is not sitting next to Danna. Lana is sitting to Danna's right, but not necessarily beside her. Which of the five girls is sitting at the far right of the bench?

Problem 5.

On a board are written all the natural numbers from 1 to 2006. First underline all the numbers divisible by 2, then all those divisible by 3 and finally all those divisible by 4. Some numbers will be underlined several times. How many numbers will be underlined exactly 2 times?

Answer Sheet

	Comment	Answer
1		
2		
3		
4		
5		

Score

1	2	3	4	5	Total

Middle School

Mock Exam 1

(Middle School)

Name: ______________________________ Grade: __________

Date: ______________ Start Time: __________ End Time: __________

Problem 1.
How many minutes does half of a third of a quarter of a day last?

Problem 2.

The figure shows a cube whose edge measures 12 *cm*. An ant moves on the surface of the cube from vertex K to vertex G along the trajectory shown in the figure. The length of the path made by the ant is

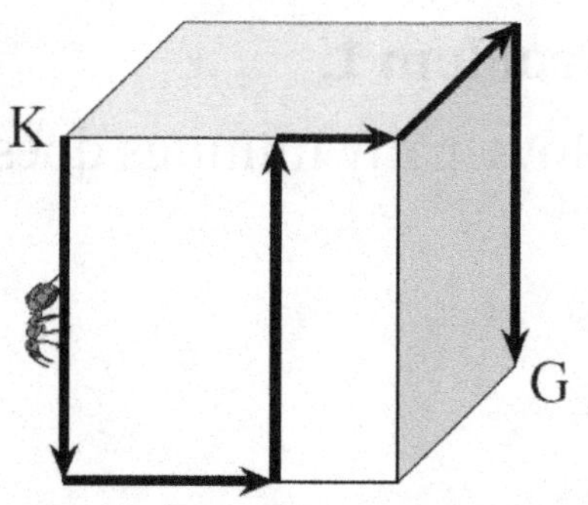

Problem 3.

Amy cuts a sheet of paper into 10 pieces. She then she takes one of these pieces and cuts it again into 10 pieces and goes on like this three more times (i.e. 5 times in total). How many pieces of paper does she end up with?

Problem 4.

50% of the students of a School have a bicycle. 30% of them (of those who have a bicycle) have rollerblades. What percentage of the students have both bikes and rollerblades?

Problem 5.

A train leaves from the terminal station with 114 people on board. At each stop 13 passengers get off and 6 get on. After how many stops is the number of passengers on board as close as possible to half of the initial one?

Answer Sheet

	Comment	Answer
1		
2		
3		
4		
5		

Score

1	2	3	4	5	Total

Mock Exam 2
(Middle School)

Name: ______________________________ Grade: ________

Date: ____________ Start Time: ________ End Time: ________

Problem 1.
In a triangle ABC, the angle at A is three times the size of the angle at B and half the angle at C. How many degrees is the angle at A?

Problem 2.
Two girls and three boys ate a total of 16 servings of ice cream. Each boy ate twice as much as each girl. How many servings of ice cream would three girls and two boys with the same passion for ice cream?

Problem 3.

The figure opposite shows the plan of a room. The adjacent walls are perpendicular to each other. What is the area of the room?

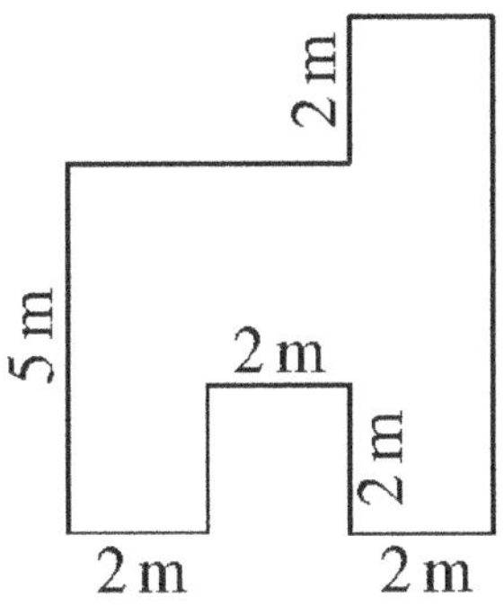

Problem 4.

The crows that live in my garden have all taken off; then each crow perched on a different pole, except one that unfortunately found no free poles. After a while they moved and are now perched on the poles in pairs and one pole is left free. How many poles are there in my garden?

Problem 5.

A set of rings are linked together as shown opposite to form a chain 1.7 m long. How many rings are needed?

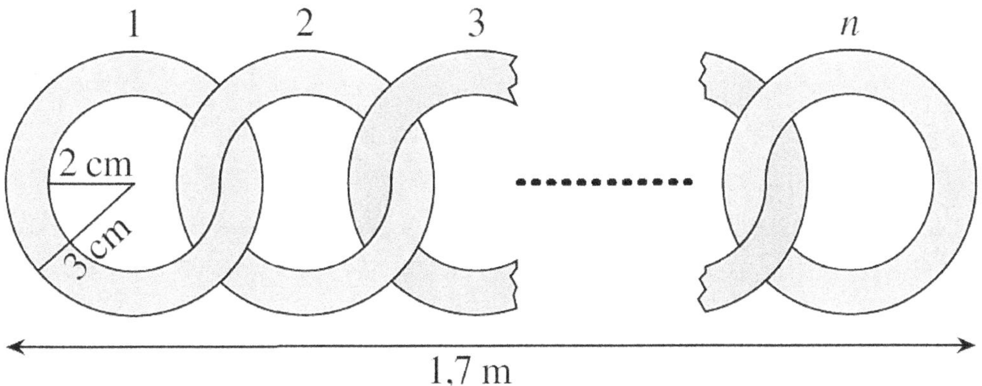

Answer Sheet

	Comment	Answer
1		
2		
3		
4		
5		

Score

1	2	3	4	5	Total

Mock Exam 3

(Middle School)

Name: ______________________________ Grade: ________

Date: ____________ Start Time: ________ End Time: ________

Problem 1.

Consider a square target as shown in the figure. The score to be obtained is inversely proportional to the area of each region. If an arrow in region B is worth 10 points, then an arrow in region C is worth:

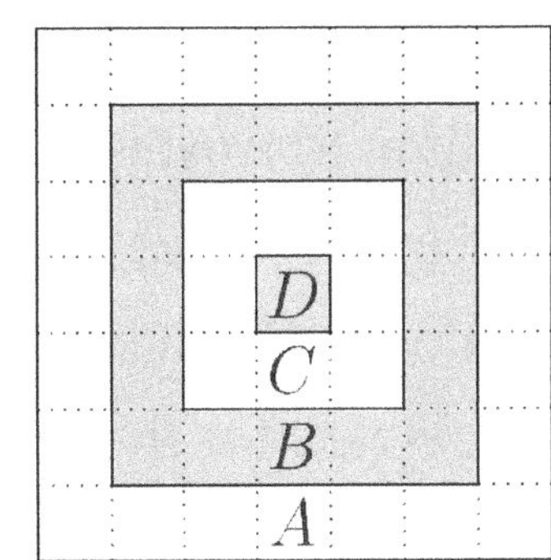

Problem 2.

A group of classmates are planning an excursion. If everyone contributed 14 dollars to travel expenses, they would have 4 dollars less than necessary; if, on the other hand, each of them contributed 16 dollars, they would have 6 dollars left over. What should be the contribution of each to collect exactly the amount necessary for the trip?

Problem 3.

On a grid like the one shown in the figure, two wires are stretched to connect node A, one with node C and the other with node B. If AC is 3 m long, how many meters is AB long?

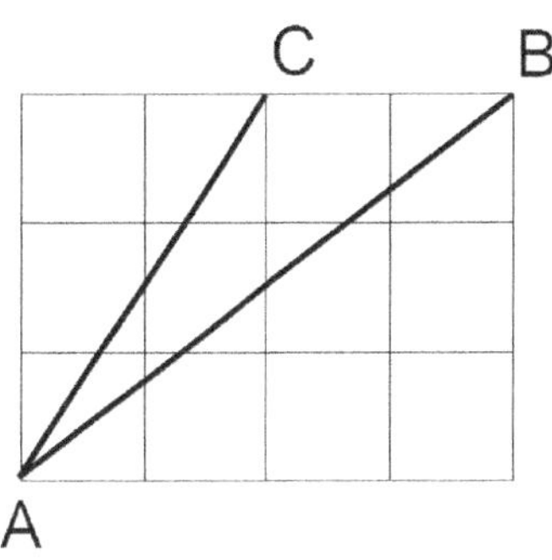

Problem 4.

A guardian works 4 consecutive days and rests on the fifth. Today is a day of rest and it is Sunday; what is the number of working days (for the guardian) that elapse from today to the first Sunday that will again be a day of rest for the guardian?

Problem 5.

In a figure skating competition, all judges are required to assign a vote in whole numbers to each competitor. The arithmetic average of the votes obtained by Mary is 5,625. At least how many judges should the jury be composed of for this result to be possible?

Answer Sheet

	Comment	Answer
1		
2		
3		
4		
5		

Score

1	2	3	4	5	Total

Mock Exam 4

(Middle School)

Name: ______________________________ Grade: ________

Date: ____________ Start Time: ________ End Time: ________

Problem 1.

Tonight it rained heavily; 20 liters of rain fell per square meter. How high is the water level in the bucket I had left, empty and not upside down, in the garden?

Problem 2.

The figure shows an equilateral triangle and a regular pentagon, partially overlapping; in particular, one of the sides of the triangle lies on the same line as one of the sides of the pentagon. How much does the angle denoted by x measure in degrees?

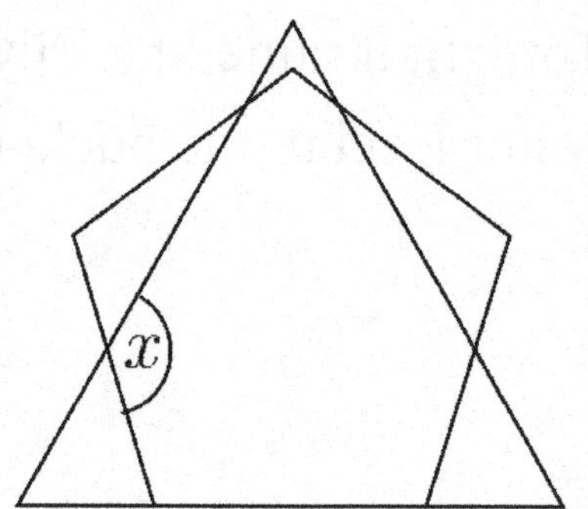

Problem 3.

The average age of the grandmother, grandfather and 7 grandchildren is 28 years. The average age of the 7 grandchildren is 15. What is the age of the grandfather if he is three years older than the grandmother?

Problem 4.
In the figure beside, the triangle is equilateral. By what number should we multiply the area of the small disk to obtain the area of the large disk?

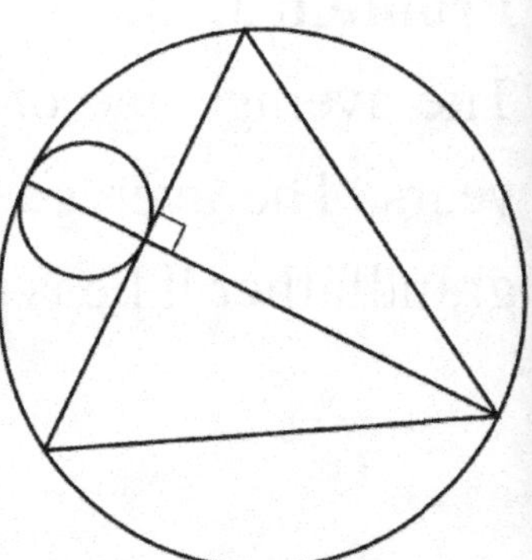

Problem 5.

A non-zero natural number is written on each side of a cube, and on each vertex we write the product of the numbers written on the three faces adjacent to this vertex. The sum of the numbers placed on the vertices of the cube is 70. What is the sum of the numbers placed on the faces of the cube?

Answer Sheet

	Comment	Answer
1		
2		
3		
4		
5		

Score

1	2	3	4	5	Total

Mock Exam 5
(Middle School)

Name: ______________________________ Grade: ________

Date: ____________ Start Time: ________ End Time: ________

Problem 1.

The figure shows the rectangles *FGHI* and *IGJK*. If *FG* and *FI* measure 4 *cm* and 3 *cm*, respectively. What is the area of the rectangle *IGJK*?

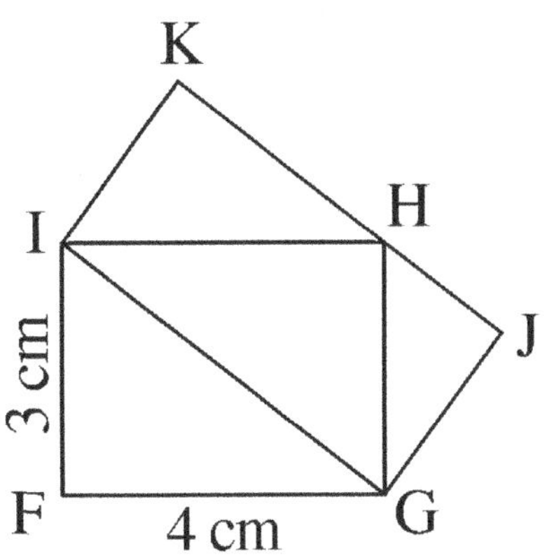

Problem 2.

The arithmetic mean of 10 different positive integers is 10. What is the maximum value of the greatest of these 10 numbers?

Problem 3.

There are 64 liters of wine in a barrel. Let us replace 16 liters of wine with 16 liters of water; let us assume that the two substances mix uniformly and that the volume of the mixture is the sum of the two volumes. Now replace 16 liters of the mixture with 16 liters of water; we wait for the two substances to mix and repeat the operation one more time. In the end, how many liters of wine (obviously mixed with water) remain in the barrel?

Problem 4.

One of the numbers below is the result of the operation:
$M = 298 \times 333 \times 352 \times 710 \times 743 \times 745$. Which one?

(A) 13727978688124860

(B) 16727978688124000

(C) 14727978688124860

(D) 13727978688124800

(E) 11727978688124800

Problem 5.

Johnny has a three-digit combination lock. He has forgotten the code, but knows that the three digits are all different and that the first is the square of the ratio of the second and third digits. How many attempts will Johnny have to make at most to open the lock?

Answer Sheet

	Comment	Answer
1		
2		
3		
4		
5		

Score

1	2	3	4	5	Total

Mock Exam 6

(Middle School)

Name: ________________________________ Grade: __________

Date: ______________ Start Time: __________ End Time: __________

Problem 1.

Calculate

$$\left(1-\frac{1}{2}\right)\left(1-\frac{1}{3}\right)\left(1-\frac{1}{4}\right)\left(1-\frac{1}{5}\right)+0.8$$

Problem 2.

We roll a die (not rigged), with faces numbered 1 to 6. Which of the following events is the most likely? The output of a number: less than 5, divisible by 3, even, greater than 3 or odd.

Problem 3.

There were 5 budgies in a cage. Their average price was 60 dollars. One day during the cleaning of the cage the most beautiful flew away. The average price of the rest is 50 dollars. What was the price of the budgie that escaped?

Problem 4.

A grandmother tells her grandchildren: "If I made 2 cakes for each of you, I would have enough dough left to make exactly 3 other cakes. However, I cannot make 3 cakes for each of you, as I would not have the dough for the last 2 cakes". How many grandchildren does that grandmother have?

Problem 5.

The figure shows a graph paper on which a triangle is drawn. If the side of each square is 1 *cm*, what is the area of the triangle (in square centimeters)?

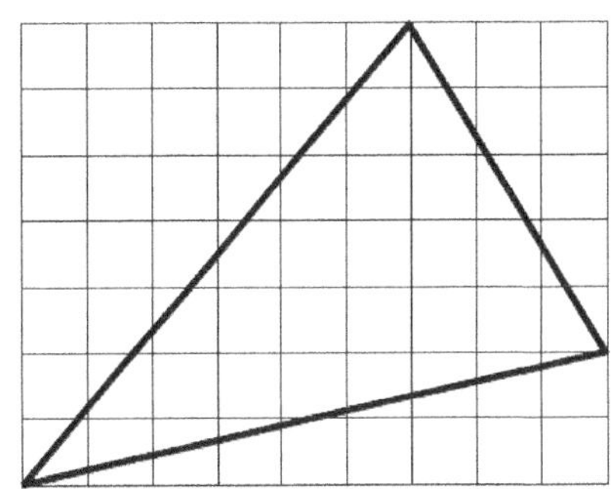

Answer Sheet

	Comment	Answer
1		
2		
3		
4		
5		

Score

1	2	3	4	5	Total

Mock Exam 7
(Middle School)

Name: ______________________________ Grade: ________

Date: ____________ Start Time: ________ End Time: ________

Problem 1.

A 1/3 liter bottle is 3/4 full. How many centiliters of liquid will it contain after pouring 20 centiliters into a glass?

Problem 2.

A survey of 3507 students in New Jersey revealed that last year 2500 of them took part in the "MOEMS" mathematical competition and 1400 in the "Young writers" literary competition. If only 7 of the students interviewed did not participate in any competition, how many participated in both competitions?

Problem 3.

The solid shown in the figure is made up of two cubes. The smallest, whose side is 1 *cm* long, is placed to the right side of the largest cube whose side is 3 *cm*. What is the total surface area of the solid, in square centimeters?

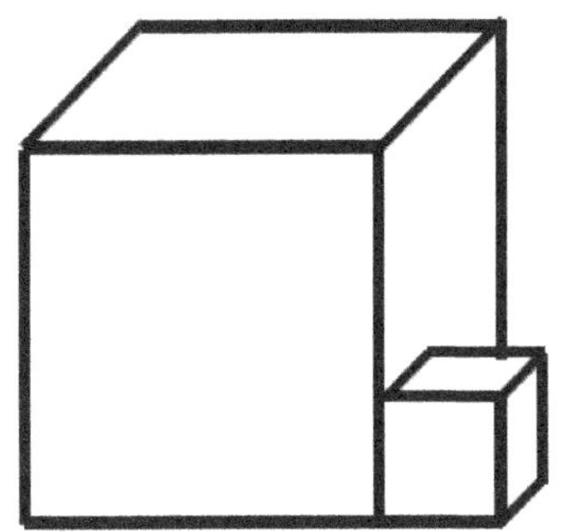

Problem 4.

If I give James two chocolate bars, he lends me his bike for 3 hours. If I give him 12 candies, he lends me his bike for 2 hours. Tomorrow I'll give him a chocolate bar and 3 candies. He will lend me his bike for:

Problem 5.

There are 21 students in a class and no two girls in it are friends of the same number of boys in the class. How many girls can be in that class at most?

Answer Sheet

	Comment	Answer
1		
2		
3		
4		
5		

Score

1	2	3	4	5	Total

Mock Exam 8
(Middle School)

Name: ______________________________ Grade: ________

Date: ____________ Start Time: ________ End Time: ________

Problem 1.

My eraser is of one color only: if it is blue, it is round; if it is square, it is red; it is blue or yellow; if it is yellow, it is square; it is square or round. How is my eraser?

Problem 2.

Annie, Bruno, and Carl pooled their savings to buy a tent. Carl contributed 60% of the price, Annie 40% of the unpaid. Bruno added the 30 dollars that were missing. What was the price of the tent?

Problem 3.

Two trains of the same length are traveling together on a double track line, the first at 100 km/h and the second at 120 km/h. When they cross each other, a passenger observes from a window of the second train that it takes exactly 6 seconds for the first train to pull out completely in front of them. How many seconds does a passenger on the first train see the second train pass in front of him?

Problem 4.

Let ABC be a triangle (not reduced to a segment) whose sides AB and AC measure 5 *cm*, and whose angle $B\hat{A}C$ measures more than $60°$. The length of its perimeter, measured in centimeters, is a whole number. How many such triangles are there?

Problem 5.

When a kangaroo pushes up with its left leg, it jumps 2 meters; and when pushing up with its right leg it jumps 4 meters; finally when pushing up with both legs, it jumps 7 meters. What is the minimum number of jumps that the kangaroo would have to perform to cover a distance of exactly 997 meters?

Answer Sheet

	Comment	Answer
1		
2		
3		
4		
5		

Score

1	2	3	4	5	Total

Mock Exam 9
(Middle School)

Name: ______________________________ Grade: ________

Date: ____________ Start Time: ________ End Time: ________

Problem 1.

A rectangle is divided into 7 squares. The side of the gray squares is 8 *cm*. How long is the side of the large white square?

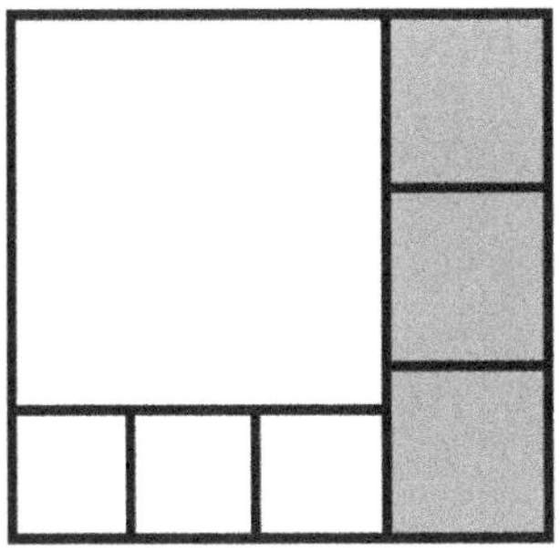

Problem 2.

How many positive integers n have the following property: of the (positive) divisors of n other than 1 and n, the largest is 15 times the smallest?

Problem 3.

How many isosceles triangles (two by two non-congruent) with area 1 have one side of length 2?

Problem 4.

In each of the vertices of the hexagon shown in the figure there was a certain number of pebbles; on each side is written the sum of the pebbles that were present in the two adjacent vertices. How much is the missing sum worth?

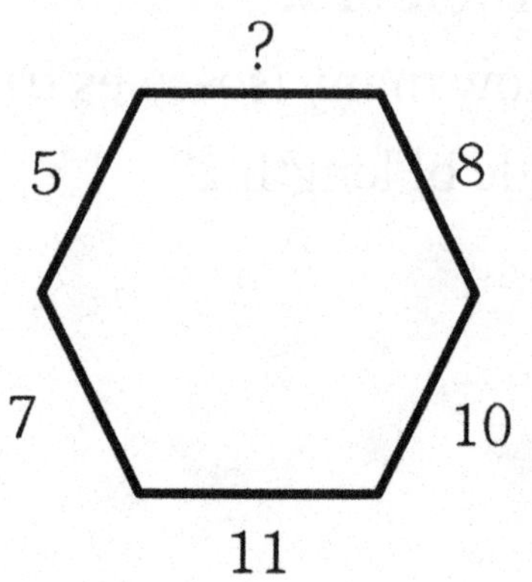

Problem 5.

The figure shows a square divided into 25 little squares, in each of which the center is marked; with a thicker line 3 obstacles are drawn. We want to go from A to B passing from one center to another only in vertical and/or horizontal direction, avoiding obstacles and by the shortest route. How many paths from A to B are there that respect all these conditions?

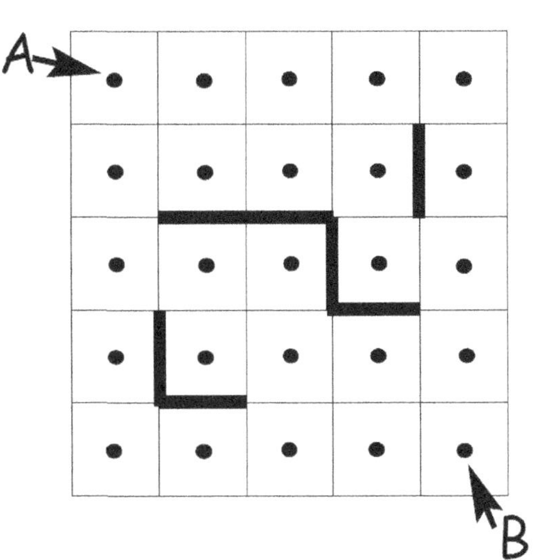

Answer Sheet

	Comment	Answer
1		
2		
3		
4		
5		

Score

1	2	3	4	5	Total

Mock Exam 10
(Middle School)

Name: ______________________________ Grade: ________

Date: ____________ Start Time: ________ End Time: ________

Problem 1.

Let $Z(n)$ be the sum of the odd digits of the number n. For example, $Z(5) = 5$, $Z(8) = 0$ and $Z(5683) = 8$. How much is $Z(1) + Z(2) + Z(3) + \cdots + Z(100)$ worth?

Problem 2.

What is the 2004-th decimal place (i.e. after the comma) in the decima representation of the number 1/700?

Problem 3.

A locomotive pulls a train of 5 wagons: A, B, C, D and E. How many ways can the wagons be arranged if the locomotive must always be closer to wagon A than to wagon B?

Problem 4.

The figure shows 4 partially overlapping squares with 11 *cm*, 9 *cm*, 7 *cm* and 5 *cm* sides. What is the difference between the area of the hatched region and the area of the gray region?

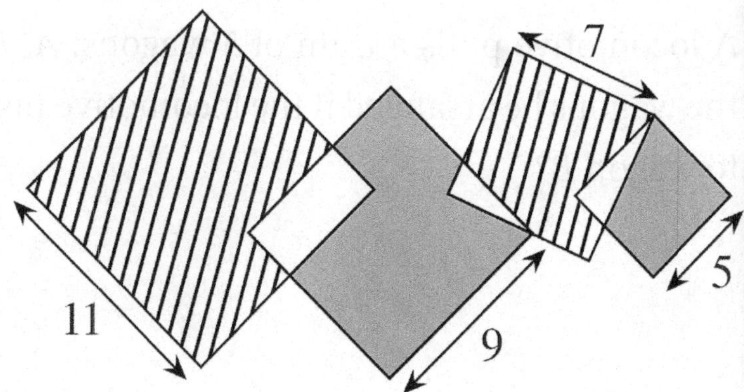

Problem 5.

Mom washed Bob's socks - 5 black pairs, 10 brown pairs and 15 gray pairs - and asked him to put them away after matching them by color. Bob instead put them all mixed up in a box. Now Bob has to go on a 7-day camping trip: what is the minimum number of socks that he just needs to take out of the box, even in the dark, to be sure he has at least 7 pairs of well-matched socks?

Answer Sheet

	Comment	Answer
1		
2		
3		
4		
5		

Score

1	2	3	4	5	Total

Summary of Results

	Mock Exam	Elementary School		Middle School	
		Partial Score	Final Score	Partial Score	Final Score
Tournament 1	1				
	2				
	3				
	4				
	5				
Tournament 2	6				
	7				
	8				
	9				
	10				

Answers

Elementary School

Mock Exam 1

1) 911	**2)** 116	**3)** 28	**4)** 15	**5)** 12 *l.*

Mock Exam 2

1) 12 *min.*	**2)** 32	**3)** 16	**4)** 644	**5)** 44

Mock Exam 3

1) 980	**2)** 80 *m.*	**3)** 24 *min.*	**4)** 12 y.o.	**5)** 900 m^2.

Mock Exam 4

1) 1 / 4	**2)** 36	**3)** 8	**4)** 19	**5)** $ 28.5

Mock Exam 5

1) 5	**2)** 45 cm^2.	**3)** 3	**4)** 667	**5)** 6

Mock Exam 6

1) 2	**2)** 11 *cm.*	**3)** 22 *dm.*	**4)** 22	**5)** 16 cm^3.

Mock Exam 7

1) 2403	**2)** 1000	**3)** 1	**4)** 4	**5)** 31

Mock Exam 8

1) 400 cm^2.	**2)** 2	**3)** 35 *dm.*	**4)** 4	**5)** OBACE

Mock Exam 9

1) 5 *cm.* **2)** 39 **3)** 6 **4)** 8 **5)** 101 *m.*

Mock Exam 10

1) 4 **2)** 12 **3)** 24 **4)** Lana **5)** 501

Middle School

Mock Exam 1

1) 60 *min.* **2)** 60 *cm.* **3)** 46 **4)** 15 % **5)** 8

Mock Exam 2

1) 54° **2)** 14 **3)** 30 m^2. **4)** 3 **5)** 42

Mock Exam 3

1) 20 **2)** $ 14.8 **3)** $15/\sqrt{3}$ **4)** 28 **5)** 8

Mock Exam 4

1) 2 *cm.* **2)** 132° **3)** 75 y.o. **4)** 16 **5)** 14

Mock Exam 5

1) 12 cm^2. **2)** 55 **3)** 27 *l.* **4)** D **5)** 4

Mock Exam 6

1) 1 **2)** < 5 **3)** $ 100 **4)** 5 **5)** 25.5 cm^2.

Mock Exam 7

1) 5 *cl.* **2)** 400 **3)** 58 cm^2. **4)** 2 *h.* **5)** 11

Mock Exam 8

1) B – R **2)** $ 125 **3)** 6 *s.* **4)** 4 **5)** 144

Mock Exam 9

1) 18 *cm.* **2)** 2 **3)** 3 **4)** 7 **5)** 12

Mock Exam 10

1) 501 **2)** 8 **3)** 60 **4)** 64 cm^2. **5)** 16

Made in the USA
Las Vegas, NV
27 February 2024